# The Moon's Nodes

## and Their Importance in Natal Astrology

### George White

Founder of the London Astrological
Research Society

Copyright 1989 by American Federation of Astrologers, Inc.

All rights reserved.

No part of this book may be reproduced or transmitted in any form or by any means, electronic or mechanical, including photocopying or recording, or by any information storage and retrieval system, without written permission from the author and publisher. Requests and inquiries may be mailed to: American Federation of Astrologers, Inc., 6535 S. Rural Road, Tempe, AZ 85283.

First Printing 1928

Current Printing 2004

ISBN: 0-86690-171-X

Published by:
American Federation of Astrologers, Inc.
6535 S. Rural Road
Tempe, Arizona 85283

Printed in the United States of America

# Foreword

In presenting this work to students, it is only just to acknowledge my indebtedness to the late Mr. Henry Duke, under whom I studied Astrology in the late seventies.

Descended from a line of astrologers, he was a storehouse of knowledge, much of which seems lost to the present generation. He directed my attention to the Moon's Nodes and other matters of great astrological import, and I am sure no one would be more glad of the success that I have achieved in pursuing researches on these little known lines than he. I desire in this book, and one to follow it, to keep his memory green, as well as add my moiety to the advancement of knowledge in this the most religious of the sciences.

I especially desire to acknowledge the assistance I have received from Miss A. Geary, the Hon. Secretary of the London Astrological Research Society, who has acted as my amanuensis. Were it not for her generous and timely aid, this book could not have been produced at the present time.

*G. White*
*London, 1927*

# Contents

| | |
|---|---|
| Chapter I. Introduction | 3 |
| Chapter II. What Is Meant by the Moon's Nodes | 9 |
| Chapter III. How the Position of the Nodes in the Horoscope Affects the Stature of the Native | 13 |
| Chapter IV. The Part the Nodes Play in the Popularity, Success and Fame of an Individual | 17 |
| Chapter V. Nodal Epochs in Life | 27 |
| Chapter VI. The Moon's Nodes in Relation to the Signs and Houses | 35 |
| Chapter VII. Eclipses Generally, and Their Effects by Direction in Horoscopes | 51 |
| Chapter VIII. The Lights Individually and the Nodes | 67 |
| Chapter IX. The Moon's Nodes in Relation to the Planets | 71 |
| Chapter X. Conclusion | 85 |
| Maps and Explanatory Tables | 93 |

# The Moon's Nodes

# The Moon's Nodes

## Chapter I

# Introduction

A KNOWLEDGE of the Moon's Nodes has existed from very ancient times. That is apparent from the known fact that our predecessors of thousands of years ago were able to forecast Eclipses. They were not only acquainted with the Nodal points, but seem very early to have noticed that those points conveyed indications or warnings as to the dealing of the Higher Powers with mankind. The names they attached to these points were accordingly somewhat striking.

In their Cosmogony they pictured something they called a Dragon, stretched across the heavens; and ascribed to this Dragon an over-shadowing effect upon the affairs of mortals. (The terrestrial one killed by St. George is presumably a comparatively puerile descendant of this idea.)

This Celestial Dragon was their way of depicting by a sort of hieroglyphic, a separate influence—or pair of influences—apparently resident in the heavens.

They had already noted the effects of the Sun, Moon and Planets. The Zodiac and what it implied was already known to the wise. But here were points showing strong separate influences, which, like the divisions of the Zodiac, needed picturing; and with something of the inspired skill that had pictured the divisions of the Zodiac, they inscribed across their map of the heavens this monster reptile. That they did not much err in judgment with regard to this disturber of the course of things, the intelligent reader will later be convinced.

Not only is it the races whose astrological notings have come to us largely through the writings of the immortal Ptolemy, that had made the discovery of these strange influences—influences that were indicated by neither Star nor Planet—but which yet seemed to interject themselves and exercise power independent of the objective heavenly bodies, and often apparently at variance with them—but the whole of the East seem to have noted the astrological value of these points, and a monster called the Serpent or the Dragon, with little variation of form, was used to depict the general influence of these points. Hindu astrologers use the symbol to this day; the points they call Ratu and Ketu respectively. Moreover, they value them as portents much more strongly than anything in the records which have descended to us through Mediterranean sources. In fact, Hindu astrologers usually treat them as the most potent influence in the affairs of man.

But nothing very specific with regard to the natures of these points has been handed down to us in the West. All that has been published with regard to them (in this country at all events) could be written in a few sentences at the most; but by writing this book I intend to enable anyone who is capable of examining his or her horoscope, to understand their influence both in the Radix and in the progressed figure. As to the importance of the Nodes, I may add that no one can live to 40 years of age without experiencing their favour or their malice, and in due course of time, both!

Singularly enough, some modern astrologers have not only failed to observe any effect from the Nodes, but have discouraged students from making efforts in regard to them on their own behalf. They have scouted the idea that mere mathematical points, however important, could have any effect. There was no material body there, so how could they influence? I might reply by saying the same of the

Ascendant or Midheaven, and ask as derisively, "How can they affect anything?" But I will reply differently. I would ask the objectors, "Do you suppose that the masses of silicate and metallic ores, gases, etc., which the Spectroscope reveals as the constituents of the Planets, have the life-long effects on mind and earthly destiny that you note? That our affections and our finances are affected by these mineral-like lumps? " Seeing the accordance between the heavenly movements and our mundane affairs, clearly provable as they are, is it not palpable that the apparent astrological effects of Zodiac and Planets, Ascendant and M.C. and other important points, are rather the indications given by the Cosmic Rulers, or other great forces serving under the Eternal Father? They are used for our guidance. Under what authority, or by what reasoning, I ask, can you suppose that the Great Centre of Power is confined to the use of masses of mineral-like matter in space by which to express himself? But still you allow him the use of two points—the Ascendant and Midheaven—and then say that whatever our experiences as astrologers are, the two points on behalf of which I write, and of which the ancients thought so much, must not be considered, as they have no body!

And here, reader, I would ask you to remember that whenever I refer in astrological matters to Planets or Nodes, or other astrological factors, as causing or producing effects, I use such terms conscious that their apparent powers are simply indications, and that all astrological points are but parts of the handwriting in the Cosmos. That physical bodies may assist in producing some of the physical results of which they are the portents, I admit; but they, equally with the points that have no physical matter attached, are but portents, or cosmic writings, and it is our duty to attend to them. The only way of knowing- their import and how to interpret them aright is by studying their concurrent indications on purely Baconian lines.

# 8 The Moon's Nodes

## Chapter II

# What Is Meant by the Moon's Nodes

MANY STUDENTS that are skilled and clever in interpreting the positions that they cull from the ephemeris are not overstocked with the astronomical outlines upon which they are built. To some of them any reference to the Moon's Nodes seems dark and puzzling. It is for the benefit of this class of reader that this chapter is written. I can commence by assuring them that there is really nothing very difficult to understand; though it is unfortunately a fact that certain textbook writers, not having much, and in some eases having no knowledge of their astrological utility, have not thought fit to burden their pages with any description of them. No doubt the strange sounding names of "Dragon's Head" and "Dragon's Tail" have contributed somewhat to this neglect.

In my various lectures upon this matter, I have usually quoted the stereotyped explanation that the Moon's Nodes are the points where the plane of the Moon's circuit cuts the plane of the Sun's circle or ecliptic. Where this is sufficient, it is sufficient; but after the lectures I found it had not proved clear to some. I think the following will make it simple.

The Moon's circuit round the Earth is nearly the same as the Sun's apparent circuit round us, but not quite. The two circles are inclined, or oblique to each other; but only to an average extent of a little over five degrees; hence, as the Moon performs its circle about once in every month, it must cross the circle of the Sun twice in that time—at one time going to the north of it and at the other to the south. The first-named crossing point is called the North Node, or Dragon's Head, and its return is made about a fortnight later to south latitude, the point of crossing being called the South Node or Dragon's Tail. In the ephemeris you will find a small column headed "Moon's Node," giving the longitudinal position and the symbol of the North Node, or Dragon's Head. The South Node, or Dragon's Tail, is the point exactly opposite in the Zodiac. At the date I am writing this, in the ephemeris the North Node is given as in the last degree of Cancer; the South Node, or Dragon's Tail, therefore, would be in the last degree of Capricorn.

I may mention here that all the Planets have their Nodes, but unless specified otherwise, when speaking of Nodes, the Moon's Nodes are always meant.

Unlike the Nodes of the Planets, which are nearly stationary, the Moon's Nodes always regress in motion; that is to say, they move backwards in the signs, their rate of regression being about three minutes per day or 19 or 20 degrees per year. These are the points near which eclipses take place, the Lights being in line with one another when passing them.

It would be well if students noted that whenever the Moon's latitude is at zero, its longitude is identical with that of the Node, for it is crossing the ecliptic.

Tables for computing the position of the Nodes at any time, will be found at the end of the book.

## Chapter III

# How the Position of the Nodes in the Horoscope Affects the Stature of the Native

BEFORE GOING into the effects which the Nodes have on the more general affairs of life, the following is an illustration of their influence on the individual, which can be easily proved by the student in studying his own horoscope and those of his friends. The position of the Nodes in the horoscope and their relation to the Ascendant have a great deal to do with the height of the native. The North Node rising close to the Ascendant, that is to say within a few degrees either side of the Ascendant, always gives height and generally a tendency to lankiness. Should the North Node be uppermost in the horoscope and in good aspect to the Ascendant, it has the same, or nearly the same effect as it would have if it were rising. If it is placed in any part of the map from the middle of the first house over the zenith round to the eighth house, there is a notable effect on stature. Should it form an evil direction with the Ascendant, the effect upon height of its being uppermost in the horoscope, will be almost if not entirely neutralised.

As regards the South Node, the converse is true. If it is near the Ascendant at the time of birth, unaccompanied by a planet or the Moon, the native will be short, and perhaps dwarfish. Its elevation from the middle of the first house round to the eighth has a notable tendency that way. Elevated and in bad aspect to the Ascendant it gives shortness of stature. Elevated and in good aspect to the Ascendant the effect is neutralised.

As an example of the effect on the stature of an individual when the North Node is rising, I would cite the horoscope of Lord Balfour. He has 26 Virgo rising and the North Node in 24 Virgo. He is extremely tall and lanky. Lord Birkenhead has the Dragon's Head elevated in the eleventh house and sextile to the Ascendant. As everyone knows, he is a very tall man.

The South Node rising has the opposite effect. This is shown in the horoscope of N. G. W. Winner, a brilliant entertainer (see "Notable Nativities" 064). This gentleman had 8 Aquarius rising and the Dragon's Tail in 8 Aquarius 25. His height was 36 ins.

Another example is that of Marshall P. Wilder, the rioted dwarf humorist and author of "People I have smiled with." He has 16 Scorpio rising and the Dragon's Tail in the second half of the ninth house, two degrees from the square of the Ascendant.

Of course the sign rising has an influence in the matter also. The fixed signs and Aries give the greatest extremes in height or dwarfishness. In this connection, however, Scorpio is the least affected of the fixed signs by the Nodes being placed in it. But Scorpio on the Ascendant is quite as much affected in the matter by the elevation and aspects of them as any other sign.

Should either Node when upon the Ascendant be accompanied by a planet, I have found there is an increase in the stature, compared with what we might expect if no planet accompanied it. A good example of this is the case of Miss Zena Dare, as given in "Notable Nativities" 633. This lady has 13 Pisces rising with the Tail in 18 Pisces, and in the natural order of things would have been very short. Venus, however, was close at hand, being situated in the seventh degree of Pisces, and mitigated this.

Another factor which must be taken into consideration is where the Moon is in conjunction with one of the Nodes. For instance, should the North Node be uppermost in the horoscope and the Moon joined with either Node, the height will be increased, and in the event of the Light throwing a good aspect to the Ascendant at the same time—as in the case of the late Prince Bismarck, who had the North Node uppermost, sextile the Ascendant, while the South Node was in the fifth house joined with the Moon—the height becomes very great. On the other hand, should the South Node be uppermost and the Moon joined with either Node the stature is lessened.

Several of the horoscopes quoted further on, and for the purpose of bringing out other points in connection with the Nodes, will also illustrate their effect upon the height of individuals.

## Chapter IV

# The Part the Nodes Play in the Popularity, Success and Fame of an Individual

I NOW propose to show the effects of the Nodes when situated near the Midheaven or in aspect thereto. Advancement, promotion, applause and fame—in fact, all that makes for success—come from the Head being situated in or throwing favourable aspects to the Midheaven and tenth house generally; while failure, remorse, public contumely and disgrace come from the evil influence of the Dragon's Tail there. I think when you have considered the proof of their action or influence, which I shall give presently, you will no more dream of leaving them out of a map than you would of leaving out a planet.

Either Node placed in any part of the tenth

house, or even 10 or 12 degrees beyond into the ninth, will affect progress all the life of an individual, and enhance reputation or bring discredit or disgrace, according to which end of the Dragon is represented. The same applies when there are good or bad aspects to the cusp of the tenth. Directions by progressed motion always produce strong results, likely to last for years, but they are not generally sudden in their effects.

Each aspect formed by a Node has a complimentary aspect of the same nature formed by the other Node, be it good or bad. And as the Angles are particularly susceptible to their influences, you may accordingly say that there are no weak aspects; even a semisquare is powerful.

But while all aspects of the Nodes to the Angles are necessarily strong owing to both ends of the Dragon casting an aspect of the same nature, there is one position that needs a special word. When they lie so that one end is at 30 degrees from an Angle, the position then formed gives the semi-sextile and quincunx aspects to the Angle affected. The first of these is only slight in power and good; the other end is casting a stronger aspect and of a mixed nature, with a slight predominance of evil. You then get a strong influence of a neutral character, but still strong. The net result has the same effect as a parallel. Parallels of

Declination are important. They give the Angle paralleled or aspected as above a strength and importance greater than they otherwise would possess. Directions formed to them in course of progression act with more effect than would otherwise be the case, and the Angle is generally dignified or Increased in importance. It is unfortunate that most of the blank maps sold for use ignore the necessity of noting Parallels to either the Nodes or the Angles; but each are of considerable importance.

Of course, other effects produced by their positions or aspects are not as visible to the eye as those that affect the height; but they are of no less importance.

In or aspecting the Ascendant, the Nodes affect all personal matters to some extent; sometimes working with and often working against the tendencies otherwise shown in the horoscope. And all positions are to some extent subject to their overshadowing influence. In all personal matters the native whose Ascendant is favoured by the Nodes has a great advantage. Those afflicted by them stand at a disadvantage through life.

In giving illustrations of these effects, I prefer, and think it better, as far as convenient, to make use of the horoscopes of well-known people, and so give students the opportunity of setting up the full figures, and judging for themselves. Astrological writers have often used most of the horoscopes that I quote (and large numbers of equally well known maps, which I do not quote, confirm my observations also), and it seems very strange that it should have been left to me to point out the lessons these figures teach! But I have been observing Nodal effects ever since I commenced studying astrology under the able tuition of Henry Dukes, in the later seventies. When a few years ago I commenced lecturing upon the Moon's Nodes, it was customary for modern astrologers to regard then as points without significance in our lives. Some even poured scorn upon the idea. I am happy to note that what I have taught has already borne fruit. When I have finished this book I shall have proved the Dragon to be the most potent influence in the horoscope. But this is a digression! Now for a few illustrations.

To refer once more to Lord Birkenhead, the Dragon's Head in his horoscope is elevated in the eleventh house and sextile the Ascendant; a very powerful position for it to have, indicating a tall man whose capacities will stand him in good stead. But it is near the semisquare of the Midheaven, showing that at times he will be subject to public criticism or opprobrium, and as a political free lance he got it. When by progression the North Node came to the Midheaven he became Lord

Chancellor, and I am not the only person who did not admire the selection, irrespective of his abilities. The semisquare in the radical map assures that opinion being much current.

His startling promotion to the Woolsack at such an early age cannot be adequately accounted for without the Nodes.

An eminent Victorian physician I will take next, namely, Richard Quain. His horoscope is given in Raphael's Ephemeris for 1901. He took his degree of M.D. under the direction progressed Dragon's Head conjunction Midheaven. It brought him fame and a Baronetcy. All the Planets were below the earth with the exception of the Moon, which was placed just below the cusp of the eighth house. Does this in itself look like rising to fame? Raphael points out the evidence of his ability as a doctor, but what about his great rise in life and the obtaining of a Baronetcy? I say that it was the position of the Dragon's Head in the radix, and the fact of his taking his degree of M.D. at the time of that point being by progression close to the M.C. which brought him the honour. The planets of themselves would not have brought the recognition he deserved; their positions were against a rise to fame.

The late Mr. Cross (Raphael) also gives in his

1899 Ephemeris the horoscope of King George of Greece. The Dragon's Tail should be inserted in the tenth house in 14 Taurus. This makes it three degrees from the square of the Ascendant. Most of my readers are no doubt old enough to remember his fatuous attack on Turkey in 1897. It was in opposition to the advice of the friendly powers, and ended in the almost comic, but at the same time very tragic night stampede of his Army, which placed him in the ludicrous position of having to appeal to the friends whose advice he had flouted.

In the 1922 Ephemeris the late Mr. Cross (Raphael) took occasion to exhibit his wit and raillery at the idea of the Moon's Nodes having any influence in horoscopes. Facing his remarks upon what he called "tomfoolery" is the horoscope and portrait of President Harding, and I shall show you that he leaves out very essential parts of the map; for, good as the horoscope is in some ways, does it really account for the greatest honour that America can give to her citizens being bestowed on him, namely, that of becoming president? Raphael not only leaves out Fortuna (with which I have nothing to do now), but also these important points. If we pencil in the position of the Dragon's Head, 20 Libra, we see that it is one degree from the sextile of the Midheaven; and when Harding was elected to the presidency, the progressed Midheaven was trine the Node. This looks as though it had some influence in the matter!

As one set of publicly known horoscopes answers my purpose as well as another, I may as well keep to horoscopes published by the late Raphael, where they are not speculative or doubtful ones, and are of mature people about whom enough is known.

Lord Haig's horoscope is another interesting example of the consistency of the Nodes. From his map, given in Raphael's Ephemeris for 1918, it will be seen that 14 Capricorn 22 is rising, and I find the Dragon's Head is in the same degree. No wonder that even in his comparatively unknown days his brother officers dubbed him "Lucky Haig"!

Mr. Lloyd George has the Dragon's Head inside the Midheaven, which by its position is an asset for life, and a sign of money and high position. Moreover, the progressed figure at the time of his becoming Premier, shows the Midheaven sextile the Dragon's Head.

In "Notable Nativities" (No. 247), the horoscope is given of a young artist named "Spare," who exhibited at the Royal Academy when only 17 years of age. The Dragon's Head is just past the cusp of the Midheaven.

The late Queen Victoria came to the throne

about a year before the Dragon's Head came to the sextile of the Midheaven—a splendid start—and was at the height of her popularity at the time of the Diamond Jubilee, when the Head was conjunction Midheaven.

In striking contrast to the above, let us look at the horoscopes of a few who unfortunately have the Dragon's Tail uppermost in their maps, and in bad aspect to an angle.

Richard Buckham "Notable Nativities" (No. 332), executed for murder: Tail elevated square the Ascendant.

W. Horsford "Notable Nativities" (No. 764), the St. Neots poisoner: Tail in the tenth house, square Ascendant. Hanged!

Walter E. Shaw "Notable Nativities" (No. 071) convicted of matricide: Tail elevated square Ascendant.

A boy convicted of murder "Notable Nativities" (No. 080): Nodes square Ascendant from third and ninth houses.

Mrs. Maybrick "Notable Nativities" (No. 969), an unfaithful spouse convicted of poisoning her husband, had the Tail in the tenth house square the Ascendant.

Two Arctic explorers (one given in "Notable

Nativities" No. 783), one of whom went with Andree in his balloon expedition to the North Pole, and like his leader was lost. That it was only a foolish and ill-advised adventure of a man who had engaged in previous efforts with success is, I think, shown by the Nodes being near a good aspect to the Ascendant, but the position of Cauda high in the tenth predicted failure.

The other explorer is Shackleton. In his radix the Tail is semisquare the Midheaven, and had come to the conjunction of the M.C. about the time of his ill-fated expedition. We know the heroic efforts and failure of the last three years of his life. In a few years he would have had the Ascendant trine the Dragon's Head; as it was his efforts ended in failure as far as he was concerned. Many other instances of this sort could be given.

Though I have taken the above horoscopes from "Notable Nativities" for the sake of example, the reader will have to insert the Nodal points for himself in the maps. The late Alan Leo, like many other modern astrologers, was not aware of the importance of the Nodes, and therefore omitted them. The unfortunate effect of this omission is shown on referring to the map given for the radical election of the exact time when Leo decided to launch the magazine "Modern Astrology." Particulars of this map are given in "Modern

Astrology," December 1923. At first sight it would appear that the magazine would be successful and profitable, meeting with great success, but the expectations have not been fully realised, and personally I am not surprised; for if the Nodes are added to the figure, it will be found that the Dragon's Tail is in 3 Capricorn below the Ascendant. This position of the Nodes, while favourably aspecting the Midheaven and thus insuring a certain amount of honour to the, venture, militates against the growth of the magazine and its being as popular and profitable an undertaking as it would otherwise have been, being so well founded as regards other positions.

## Chapter V

# Nodal Epochs in Life

SO FAR I have only given a brief sketch with a few examples of Nodal action in its most easily traced forms. These forms have become known to quite a number of astrologers since I first made them public by lectures, and allowed a number of copies of my lectures upon "The Nodes and the Angles" to be circulated, and some copies of my "Eclipses in Horoscopes" also. It is accordingly getting quite common in some astrological circles to hear the old-fashioned names of them being used: "Caput" for the Dragon's Head or North Node, and "Cauda" for the Tail or South Node.

I have frequently been requested to enlarge upon the subject and also to publish the information that I have been able to dispense in my various other lectures upon the several manners and phases of action portrayed by this over-shadowing influence; I do so willingly.

In these high latitudes it is only when one of the Equinoctial points (0 Aries and 0 Libra) are rising that the Nodes, by lying upon one pair of the Angles of the figure, are squaring the other pair of Angles.

Near the Equator the parts of the Zodiac upon the Angles are almost uniformly in square to one another; but as we recede from the Equator this uniformity is broken up and we approach latitudes in which while a Node is conjunction one Angle it may be either sextile or trine another, providing that no equinoctial point is rising at the time. Hence we often have a double action of the Nodes, blessing one pair of Angles and afflicting the other pair, or afflicting or else blessing both, as the case may be.

I say "pair of Angles" above, and this, of course, is literally true; but it needs a small qualification which will be made obvious in a later Chapter. For the present it is sufficient to say that the effects are most noticeable upon the Ascendant and Midheaven, and what they stand for respectively, though the complimentary Angles, and what they stand for, are affected also.

It will have been gathered from what I have already said that the Nodes aspecting the Ascendant and Midheaven are—at least at times—as important as if actually occupying the Angles. That is so, but allowance has to be made many ways, and in some horoscopes that I have studied, the Nodes well or badly aspecting seem to have proved themselves stronger than one would expect from their simply resting upon these points. Moreover, there is the added complexity of the House positions and sign positions from which they are operating. Many of the most fortunate people to be found are those who have the Nodes serving them well by favourable aspect to both Angles. No horoscope will prove very satisfactory, whatever the disposition of the Planets, if the Nodes are seriously afflicting both the Ascendant and Midheaven. There appears to be a more easily discernible difference in this respect in regard to their influence upon the Ascendant than there is with regard to the Midheaven. That, of course, might be expected. The Ascendant stands for the intimately personal and it is not only in height that their influence is felt. Personal characteristics are affected in ways readily noticeable, but that matter will come more naturally in place in the Chapter upon the Nodes and the Houses. I think, therefore, in the present chapter it is best to confine myself to Aspects to Angles, as such, though of course the reader will recognise that the House positions they hold in the horoscope have their effects in shaping the results.

To illustrate this I would refer to the horoscope of the late Lord Beaconsfield. A certain noted astrologer (the late Raphael) has said that he could not see enough in this horoscope to account for his rising from a relatively low position to the high estate he reached. Quite so, there is not enough in the parts of the horoscope that Raphael limits himself to; but once add the Nodes and you have the missing quantity. In Lord Beaconsfield's radix you have the Dragon's Head sextile to but below the Ascendant, and within five degrees of the trine of the Midheaven. D'Israeli became Prime Minister in 1868 with the North Node just past the junction of the Ascendant, and the sextile to the Midheaven in full operation.

Another instance is that of General Sir R. T. Baden-Powell ("Notable Nativities" No. 837). He has the North Node on the cusp of the second house, sextile the Ascendant, and trine the Midheaven.

As this horoscope is also a fine example of two other forms of Nodal activity, it is set out in full at the end of the book, and students are advised to study it carefully. The other forms of Nodal action there shown will be dealt with under their own proper headings.

As compared with these examples take the case of Gandhi, who has the North Node close to the Midheaven and square the Ascendant. The agitation which he aroused in India he admits is a failure, or partly so; and it certainly has entailed much suffering and imprisonment upon him. But his reputation in India is such that millions of the inhabitants look upon him in almost the spirit of worship. He probably has more personal admirers than has any other living man today.

This is an instance showing the cross workings of fate where the Angles are favourably and unfavourably affected at the same time.

We will now consider the effect of mixed aspects and directions in a progressed horoscope, and as an example I will take the map of the late Mr. Gladstone, who was born on December 29, 1809. His figure shows 9 Capricorn rising and 17 Scorpio culminating. The North Node is in 20 Libra (the same position as it occupied in the map of President Harding) and as it is uppermost, he was tall. In 1843, Mr. Gladstone took Cabinet rank with Dragon's Head sextile Midheaven, was accordingly a great success and became very popular. Twenty-five years afterwards he became Prime Minister for the first time, and held office from 1868 to 1874. During that period the Nodes came to the square of the Midheaven, and what occurred during those years is astrologically very instructive. The House of Commons had a very busy and tumultuous time, but the great feature to my mind was the extraordinary and malignant slanders circulated throughout the country at the time, and that not secretly. The widespread abuse heaped on Mr. Gladstone's head does not seem feasible to have taken place, and only those of us who lived through that time are likely to realise it. Secret stories and public abuse convinced religious people that Mr. Gladstone had secretly entered the Church of Rome, and was virtually an envoy of the Pope to destroy the English Church. This and much more was freely talked about, and popular preachers gave countenance to it from their pulpits. Most or all of the religious papers (then of much more consequence than now) seemed as though they believed these fantastic stories, while the Ballot Act and other activities of his brought all the propertied classes against him. Nothing more typical of the Nodes square the Midheaven could have been, and after the election of 1874, he was for the first time for 40 years out of all office, and calumny had helped to score a great victory over him for a time. Moreover, the dislike of Queen Victoria for him had become well known.

Now look at the end of his life in 1898. His immense popularity was a feature of the age, and I suppose all will recollect the solemn funeral the nation accorded him. At that time the progressed Midheaven was close to the trine of the Dragon's Head.

The Nodes in the main seem aligned toward material matters and mental processes, and the order or tendency of things toward what might be termed "luck." A tide of affairs accelerating or retarding our aims, intentions and interests. The Head blessing our endeavours, and the Tail causing difficulties or miscarriages and frequently disgrace.

Of themselves they are rarely spectacular in effect, except when placed upon the Angles, but they are always ready to enhance or thwart anything they become entangled with. Their great characteristics in action are slow, lasting and (often) overpowering.

I have (in the past) several times referred to the unmoral nature of the Nodes, but there are points of exception needed in this generalisation. Caput near the Ascendant gives a natural refinement and gentility, and in sextile or trine to either Angle gives sincerity, and so a tendency to trustworthiness, as well as making for success. When the Nodes are aspecting the Midheaven their testimony as to success or failure, renown or ignominy, seem to bear no relation to what is *deserved* by the individual. In the case of the maps of murderers, quoted earlier, who had the Tail in the Midheaven, it was the failure to achieve their ulterior purpose, rather than the nature of the purpose, that was portrayed. No doubt the Tail afflicting the point of honour had its demoralising effect upon the individual; but it is the fact of being "found out," or, at least, of their being convicted, and not the question of guilty or not guilty that is shown by the Nodes. Their efforts were failures, and punishment and disgrace fell upon them. Instead of getting betterment, which they aimed at, they reaped contumely and disgrace. Honour and success, disgrace and failure, are mainly in the hands of the Nodes, and it is most important to bear in mind that people's characters are not necessarily to be judged by them excepting as will be intimated later.

# Chapter VI

# The Moon's Nodes in Relation to the Signs and Houses

THE NODES, like the Planets, have their special points in the zodiac in which they are stronger or the reverse.

The ancients gave Scorpio as the elevation of the Dragon's Tail and Gemini as the elevation of the Head. More recent writers have assumed the Tail to be elevated in Sagittary. My own observations induce me to give the whole of the northern half of the Zodiac to the Head and the whole of the Southern half to the Tail, but this does not fully settle the matter. I am of opinion that the Nodes are stronger in the common signs in general than in the other signs, and that no regular order can yet be assigned to them as regards strength.

Gemini and Virgo are certainly very favourable to the North Node, while the South Node seems less prone to mischief if placed in Sagittary or Pisces. Probably the idea of the South Node being elevated or dignified in Scorpio arose from the similarity of character of the Dragon's Tail and the sign Scorpio, each being by nature damp, low-lying and having a tendency to be evil smelling.

It is essential to recollect the nature of the Nodes and their affinity to the signs they are in. The North Node is by nature high and dry; it represents the hilltop while the South Node represents the evil smelling swamp.

It may be mentioned that the sign Pisces seems to modify the influence of both Head and Tail; that is to say, the influence of the Head is not nearly so good if placed in Pisces and the effects of the Tail are not nearly so evil when placed in this sign. I cannot say why this is so, but my experience has proved it to be the case. It may be asked, is the influence of the Nodes reversed in horoscopes for southern latitudes? I have investigated and made enquiries on this point and the testimony seems to bear out that there is no change in their nature; when in such position the Head does not take on the nature of the Tail nor the Tail that of the Head.

Now as regards the position in the Houses, I

have already partly dealt with their effects when placed in the first and tenth houses. The Dragon's Head in the first house is a position of considerable value in itself. It generally shows a person who will go upwards in life. Either Node in the first house tends to make a person conspicuous in other matters than height. These natives (to use an old astrological term) are generally people who attract some attention and have a good conceit of themselves whether the rising Node is Caput or Cauda, and whether they be tall or short.

With Caput ascending, there is generally some real ability, and the person who has this will attract respectful attention.

Those with Cauda ascending are less fortunate and sometimes only command amused attention; still, they are not likely to be passed by lightly.

A lordliness often accompanies the Dragon's Head in, or just above, the first house, which usually appears proper and in place. The Tail also gives a sense of importance, but frequently with a ludicrous appearance, and is often designated "Cheek." But these less fortunate ones often fare pretty well in the world, if helped by planetary positions; and they generally have the necessary self-confidence to help them on. They seem lacking, however, in the natural refinement and self-uplift possessed by Caput people; they are usually less scholastically inclined, and suffer accordingly. I will cite the case of a man I have known from childhood, which is, I think, a fairly typical Cauda case. At school he was a perfect dunce, and accordingly he was looked upon as a fool. On leaving school and going to practical work he soon showed some ability. He not only developed general mechanical skill, but in time became a workman much above the average. He has left school well over 60 years now and has been a credit to those who have employed him. If he had not had capacity, his appearance and speech would have made getting a living a very difficult task. He is now past doing manual work himself, but is a valuable assistant and overseer in the firm of which his eldest son is the senior partner! Yet this man is a dwarf, or nearly so; at school he was a dunce, and he has the Cauda rising defect of talking, even now, the wretched back-street grammar of his illiterate parents. Modern eugenists would have classed him as a degenerate unfit to propagate! But he has progeny that would be a credit to any family.

A little lady now getting close to middle age, who has the Tail about two degrees below the Ascendant, is a singular contrast to Lord Balfour, whose map was quoted earlier. Both born under Virgo, the one is relatively as short as the other is tall. I expect the lady is the more consciously ambitious of the two. That would be quite in the nature of the Nodes. Caput has a tendency to take on position as quite a natural burden, while Cauda, to whom it does not naturally accrue, struggles for the position which is attained. In the present instance, considering the circumstances under which this lady started life, she has (thanks to an otherwise good nativity) done very well for a case of Cauda rising. She has all its self-assurance and perkiness, is industrious, frugal and determined to go up in the world, if possible; but she seems to have no idea of real self-improvement. Double negatives remain a considerable factor in her syntax, and all the common barbarisms of speech (which are not indictable) still mingle with merry laughter. The incongruous nature of this talk does not seem to dawn upon her; and yet she is very ambitious! So very much like Cauda rising!

These are fairly typical cases. When the Nodes are nearly horizontal, as in these cases, it is the eastern one that really carries weight. As the position changes and you get, say, down to the middle of the first house, and so the corresponding Node rises to near the middle of the seventh house, the latter becomes of some importance, though not equaling the other Node. This is strongly shown when looking into the matter of stature. When Caput is down to the middle of the first house you scarcely notice its effect upon the stature, but the effects that way are very noticeable for some distance above the cusp of the house. Also the end of the Dragon that gets into the seventh house seems to have more effect when well above the cusp than when close upon or just below it. Neither isolated in the first house seems to make for robustness.

## Second and Eighth Houses

Many of the horoscopes of very fortunate people have Caput upon the cusp of the second house. Of course, in these extreme cases (I have quite a list of them) it has also favourably aspected the Ascendant or Midheaven as well, and in several cases both angles. In ordinary cases, however, when not aspecting angles, it is still a very fine position. Cauda there an evil one; often causing loss of fortune, always causing leakage, and frequently causing great financial injury by making' credit bad. Cauda in the second prevents the trader who needs credit from obtaining it upon reasonable terms, and so often shuts his shop up. Deserving people with that position rarely meet reward unless it be fairly late in life.

People with Caput on the cusp of the second house usually have their financial credit standing high for their station in life, and take care to retain that credit. They generally get good opportunities and possibilities of a successful career fairly early in life.

Folks with the Dragon placed across these two houses are usually pretty generous and free with their money; that is, according to their means, whichever way it lies, but are not often really wasteful.

The eighth is a weaker and less important house than the second, but its elevation makes it count. Dragon's Head there will often give a legacy, or a gain from wife's estate, but with great risk of the recipient being only temporarily benefitted by it. Cauda there usually brings some trouble with and ultimate losses of legacies, if any. It also inclines to bad health. One case I know in which a legacy was entirely lost by bad health, and yet the legatee did not go on the loose. The man was a sufferer from rheumatism for some years, and endeavouring to get rid of this enemy he soon got rid of the legacy, without it leaving much benefit behind.

## Third and Ninth Houses

As we get toward these houses the mere matter of elevation becomes of still more importance. Which Node is uppermost now is becoming the dominating matter.

These houses share with the Ascendant the honour of largely displaying the mind. The Nodes occupying them increase mental activity and ability. But Caput in the ninth is much the more desirable. Let Cauda be uppermost, superstition, fear of Black Magic, dreams, apprehensions and morbidity of mind may be present, though often there is considerable ability. Great dreamers often have it so placed.

I know of horoscopes of several journalists in which one of these positions occurs. But unfortunately they are mostly, if not all, too closely mixed with planetary positions for me to use as illustrations. They are mostly cases of Cauda uppermost.

The people who stand as advocates or pioneers of unpopular sciences or doctrines often have Cauda in the ninth house. Folks whose energies and abilities would bring them to the forefront if they ran exclusively on orthodox lines are frequently handicapped by this position. Espousing the unpopular is a characteristic of not only Dragon's Tail in the ninth house but also of its being placed in either of the two following houses, namely the tenth and eleventh. Of course, this only applies to the advanced, or, as occultists say, to the "evolved" class of people. Such people do much service to the race, often largely reaping contumely in return.

The Dragon's Head is upon the wrong side of the line of advantage for them! To those people who aim solely at the material, foreign investments must be a veritable pitfall with this position. Home, as distinct from foreign, commercial travelers may reap some advantage from it.

## Fourth and Tenth Houses

I have spoken much during the last three years upon the effect of the Nodes being near to, or in aspect with, the cusp of the tenth house, and so far have dealt with them mainly in regard to the angles or framework of the figure. But there is a good deal I have not referred to, and I shall now touch on some of these points.

These two houses have much to do with family stock, and in many horoscopes must give fine scope for speculation upon that matter. But as the horoscope primarily relates to the individual, diving into the question of family stock is apt to mislead; nevertheless, the Nodes aid in this respect.

The Dragon's Head in the tenth house is an excellent testimony to success in life and the reaping of honours. Such people usually *rise above* their station. The Tail shows difficulties and disappointments, failures and often disgrace. But a few people do succeed in the world despite this handicap, though it is generally by unpopular means, or as leaders or accomplished workers, in unpopular causes. To people whose horoscopes otherwise show trouble or disaster, this position makes them disastrous indeed.

I have noticed that the Dragon's Head in the fourth house often seems to cause a man to be the father of illegitimate children. I suppose it gives an urge toward increasing family stock, while the upper, and so predominate Node, gives the colour to the transaction.

The wretched case of voluntary barrenness after going through the marriage ceremony, reported by "A Nurse" in the May 1925 number of "Modern Astrology" is a case of Dragon's Tail in the tenth house in a woman's horoscope; this depriving herself of a woman's greatest glory, a family, seems to be quite characteristic. I know of other cases of a similar, though perhaps hardly of so regrettable a nature. On the other hand, I have noticed that Caput in the tenth house often accompanies large families in a female nativity.

I have several maps of illegitimate children having the Dragon's Tail in the tenth house, and think it is rather a common attribute with them.

In some bad cases of seduction where the Dragon's Tail has been in the tenth house, and a child has followed, the matter has been quite notorious, and this result has made a hash of the mother's future life.

The fortunate lie of the Nodes in these houses is highly favourable to marriage.

## Fifth and Eleventh Houses

The Dragon's Head in the eleventh house has been the making of many notable people. I need not refer again to the case of Lord Birkenhead. To professional men and all who depend upon patronage it is a most valuable asset. In fact, it is an asset wherever it occurs—even in the map of a courtesan. Being oriental of the Midheaven, its progression is always full of promise. It is a position singularly fortunate for nearly everything except gambling, betting and "wildcat" speculation. Especially for dealing with or promoting Joint Stock Companies and such like undertakings, the Head here is invaluable, but the Tail may spell ruin (Bottomley to wit).

It is common in the horoscopes of great flirts. It is more conducive to flirting than to marriage. I have known irregular unions formed under it, and whichever end of the Dragon may be uppermost in this pair of houses, there seems to be an inclination toward the production of illegitimate children.

Cauda in the eleventh is usually a very unfortunate position. Friends are not likely to be of much assistance in life. A very few politicians do well with it, in a minor way, but they do not have Peerages thrust upon them. Being oriental of the Midheaven, there are certain to be periods of life when progression brings some disaster. Loss by and treachery of friends is a pretty sure result of the position.

One characteristic is common to the Nodes in these houses whichever way up: the danger of too free a course of life. I do not think ascetics are commonly favoured with the Dragon lying across these two houses, though I know one case of a nun being recorded, she having Cauda in the elevated position.

## Sixth and Twelfth Houses

These are rather secret and obscure houses to know much about. Not being very strong houses, I think we may put them as usually of not very much account as far as the Nodes go. To those who employ menials and the lower order of assistants, the Tail in the sixth house must often be an inconvenience. To the latter class of people themselves it must tend to wretchedness while the occupation lasts.

I do not happen to be favoured with the horoscopes of many men who have taken up Army or Navy for a livelihood with these positions occurring. In years gone by a very large percentage of those people figured at one time or another as deserters. I should think the Tail in the sixth would frequently be found characteristic of this. I throw out that as a *suggestion*, and hope one day to obtain light upon it.

The position of Caput in the twelfth is certainly one of some promise, especially to those who by business or profession find public institutions useful. Equally so to those who from necessity need the assistance or shelter of such places. Its oriental position is, of course, favourable to future developments.

To psychics it must matter a good deal. I should advise those with Cauda in the twelfth house to avoid developing psychic susceptibilities.

I have met with the horoscopes of three illegitimate children, each with the Dragon's Head in the twelfth house, but there the lack of data comes in again. I do not know what became of them or whether they were cared for (as so many such fortunately are) by suitable institutions. But I know of a case of a single woman who has had several children, and one of those children she surrendered to a home, and I believe two previous children had similarly been provided for. That mother had Caput in the twelfth house, but the child that I knew of, and who was surrendered to a Catholic home, was a very fine boy with the Dragon's Head in the fourth house.

Remembering that the twelfth is the house of secret enemies, and also of regrets and self-undoing, it certainly must be a bad place for the Tail to lie. Its elevated position and its progression upwards are certainly somewhat menacing. But against that we must remember that they are a pair of very weak houses and usually of very little importance unless tenanted, and I am not now dealing with them as tenanted houses.

The Dragon, having two ends, creates a little difficulty with some minds. It looks as though the one end was likely to contradict the other! They do not contradict, but act complementary one to the other.

Elevation is an astrological doctrine that I think all students understand. And the importance of Orientalism is equally well known. The upper always gives colour to and generally dominates the lower. Likewise the eastern is of more importance than the western. With these considerations clearly before you, what I call the "Line of Advantage" becomes plain and intelligible.

If you draw a line upon a horoscope from a point a little above the cusp of the third house to a little below the cusp of the ninth house, you have what I call the Line of Advantage. And it is a very important matter on which side of that line the Dragon's Head is placed; for whichever Node is upon the upper side of that line largely dominates the map.

In practice one often finds positions formed that might destroy, or partially destroy, some of the advantages derived from having the Dragon's Head on the right side of the Line of Advantage, especially so in a nocturnal map, where it may cause a South Nodal eclipse to be formed by direction where a North Nodal one otherwise would be. Of that later! But taking horoscopes in the main, the balance of advantage will rest with those which have Caput upon the side I have indicated. With females it tends towards what their friends denominate "a good marriage"! But the Dragon, viewed generally, does not seem much disposed toward making marriages.

## Chapter VII

# Eclipses Generally, and Their Effects by Direction in Horoscopes

ASTROLOGICALLY CONSIDERED, eclipses are essentially Nodal phenomena. The same might be said of them astronomically, but as it is their significance on life that I wish to deal with, I emphasise that point. Since the discovery of the progressed horoscope they assume an importance, when understood, that must have been entirely unknown to the ancients. Hence all, or nearly all, that has been handed down to us from that source relates either to their supposed effects in mundane astrology or as transits over the genethliacal figure. So that however much or little the ancients did know in regard to Nodal influence, their effect in the progressed figure seems to have been a sealed book to them; therefore we cannot look to them for guidance as to Eclipses working out in the form of directions in the progressed horoscope. It seems strange that astrologers of a later day should have so overlooked this, one of the most important things in genethlialogy. I first lectured upon this matter of Eclipses in horoscopes in November 1922, and the subject came as a surprise and quite a new thing to my audience, though several of them had spent years practising the astrological art. Since then several of my astrological friends have enthusiastically adopted my way of dealing with and reading these portents of good and ill.

The fact that the two Nodes are of opposite natures clears much mist away, and of itself gives a grand clue as to how to view these most noticeable forms of nodal influence. And I hope we shall not in future have such a statement from a leading astrological writer as, "An Eclipse must always be considered as a privation or misfortune of some sort," as though privation or misfortune was inevitable from occurrences that frequently are most valuable, and sometimes the best thing that can occur to the native. The mere dealing with Eclipses as transits will not be the limit of possibility in the future. Moreover, the notions handed down to us (and so far relied upon) will be corrected and put upon a more reasonable footing; at least, that is a large part of my object in writing this book.

## The Moon's Nodes

Upon the purely astronomical side of Eclipses little need be said. The mere mechanical phenomena are amply explained in many useful textbooks. But old students will no doubt excuse me explaining to younger ones the mere phenomena in a few words, with a few relevant facts that I consider to be worth their bearing in mind.

As previously explained, the Nodes are the points on the Ecliptic where the plane of the Moon's orbit crosses the plane of the Sun's orbit, and these planes are inclined to one another at a mean angle of little more than five degrees. Consequently it is only when the Sun and Moon are passing one another in the neighbourhood of these points that the Moon can come sufficiently between us and the Sun to obscure or partly obscure that luminary, and so cause a Solar Eclipse. A Lunar Eclipse is simply an ordinary full moon or opposition formed near these crossing points or nodes, and when this occurs, the Earth's shadow is cast upon the Moon.

If a new Moon falls within five degrees of a Node, you have either a total or annular Eclipse of the Sun. If it falls more than five degrees away and less than 18° 36' from the Nodal point, there is only a partial Eclipse. Beyond 18° 86' from the Node the parts of their respective circles then occupied are too wide apart for an Eclipse to be formed, and the Moon's shadow is cast into space and unobservable by us.

If at Full Moon the longitude of the Sun is within 12° 24' of the Node, then there is an Eclipse of the Moon, total or partial, varying according to the distance from the Node. In other words, within that distance the Moon will catch more or less of the Earth's shadow, instead of the shadow all falling into space, as is usually the case.

The frequency of Eclipses is a matter apt to be overlooked. In 70 years we live through 140 Solar and well over 100 Lunar Eclipses. They do not come in even numbers. As a sample of that I will quote that in 1886 there were only one Solar and two very pronounced Lunar Eclipses. In each of the years 1870, 1884, 1899 and 1908, there were three Solar Eclipses. In the year 1898, there were three Lunar Eclipses. The year 1917 scored the fullest number possible for one year, four Solar and three Lunar Eclipses. The mean supply works out at about two Solar and one and a half Lunar Eclipses to the year; but these eclipses are usually mere transits and must be regarded as such. Considered as such they at times are of some importance, according to what part of the horoscope they fall upon, and how aspected, provided there is something in that horoscope in the way of directions that they can well or adversely affect. In that case the first question to ask is which Node does the affecting Eclipse happen near? I notice that the writers who threaten great things from these mere transits do so in palpable ignorance of the astrological nature and effects of the Nodes upon which the eclipses they speak of depend. And when we read what is ascribed to these transitory happenings, and remember that a man of seventy has lived through about 250 of them, well there is no need for comment!

But it is not mere transit eclipses that I am concerned with; they will come in for some consideration later on. It is those that are developed in the course of progression of the horoscope that I am now concerned with, and which form the most powerful and far reaching directions that occur, whether they be for good or ill.

They do not come in all horoscopes. The chances of getting a Solar Eclipse by direction in a life of 70 years is about two to five; and the chances of getting a Lunar one about two-thirds of that amount. But as a Lunar Eclipse frequently happens in about a fortnight before or after a Solar one, those who develop one sort in their nativity have a good chance of developing one of the other sort. Moreover the Solar variety occasionally occurs at successive lunations. For this to happen the Sun must be a good way from the

Node at the first lunation in order to be within the necessary limit of 18 degrees at the second one. Hence neither can be classed as of very first-rate importance, but by falling out as directions each will have very noticeable effects.

## Eclipses in Horoscopes

I will now deal with this class of directions and their nature, and will give illustrations, as well as statements, as to their effects. I will commence with a brief reference to what I consider the common sense of the matter.

All lunations are important when occurring as directions made by the progression of the horoscope. But the conjunction of the Sun and Moon when not near a Node is some distance apart in latitude, and occasionally as much as five and a half degrees if it happens to occur in square to the Nodes. The student must remember that when more than one degree apart in declination it loses the addition of the great power attached to a parallel of declination, a power that is equal to that of a conjunction. The Node of itself is the most potent influence in the heavens, actually stronger than a major planet, and being with this close conjunction and parallel adds immensely to its powers and moreover qualifies them; thus we have an unusually strong configuration formed for good or evil. Should it occur at the North Node or Dragon's Head, its tendency is for great good, unless it be in opposition or square to an important point. Should it occur at that evil point, the South Node, or Dragon's Tail, the effects may be disastrous.

I will here ask the student to bear in mind that Eclipses, being Nodal phenomena, the causes of the events they portend will usually be a long time in the making. The results of them will last for years. Also that they are rarely spectacular in effect, excepting when joining forces with other positions or directions, when the result may come unexpectedly into prominence; a thing that frequently happens with regard to the eyesight, or rather with the advent of eyesight troubles in the form of blindness, as I shall show later on. But though usually signaling a complete, or at least a partial change, in the course of life, it will generally be found difficult to fix a date from observed facts in life from which it can be considered that the cause of the change operated. Unless other directions in the natus are working with it, and so give a sort of starting point, the event or change will simply come about as quite a natural and ordained affair, culminating according to Lunar directions, transits, etc., with effects lasting for years. Hence the necessity of closely watching this powerful form of nodal action. And yet I am not aware of a single book on astrology that makes any reference to such a class of directions occurring! Yet in dealing with horoscopes in which they occur, both students and practitioners must often have been baffled for want of light upon this subject. Whatever point of the horoscope an eclipse falls upon will be impressed with the nature of the eclipse, whether good or bad. If a North Nodal eclipse it will be good; if South Nodal, bad as far as that particular point goes. For instance, it may fall in some part of the fifth or sixth houses, and affairs connected with these houses will be well or badly affected according to the nature of the Eclipse. Should it fall on the angle of the first or tenth house, the effects will be powerful, good or evil, according to which Node it may be connected with; but in any case any Eclipse afflicting an angle by aspect as it is formed will have a very powerful adverse effect, so that an Eclipse may really partake of two natures, beneficial so far as the local point is concerned, but very disastrous from its aspects to an angle. I think that in a general way the North Nodal eclipses are stronger in their effects than the South Nodal ones, and their good aspects to an angle are of greater importance for good. The effects of the South Nodal eclipses when in good aspect to angles, or otherwise, promise some degree of benefit, and generally of a nature not overly creditable in the details of their working out in life. The benefits from a North Nodal eclipse are not only more powerful but usually of a kind creditable to the native. I am, of course, referring to their action when simple and not complicated by the action of any planet. Falling upon the place of the Moon they will, of course, act in accordance with their natures for good or evil, and upon the Hyleg may have good or very disastrous results upon the health or life of the native. Falling upon planets, it brings the planet and whatever it stands for into great prominence for the next few years, the nature of the influence will be judged similarly from aspects to planets; likewise, gain and loss must be judged in the same manner as aspects to angles, while the house the aspects fall in will show the direction in which the aspect will work out.

Lunar Eclipses, unlike Solar Eclipses, can never be really good. Fortunately their effects are very much weaker in character, but they vary according to how the Lights lie with the Nodes. If they occur when the Moon is with the North Node they are not really as a rule of much effect, except as a cause of change; but when the Moon is with the South Node, it makes a particularly poisonous point in the horoscope, and the house in which it falls will suffer much. Such Eclipses are particularly dangerous where the sight is at all threatened, as indeed all eclipses are, but I shall make some special comments later upon this point. Falling upon angles, this form of

Eclipse is noticeable for its effect upon the health, and especially for bereavements. For instance, a native may lose a mother or a wife. There is a curious case given in "Notable Nativities" No. 299. It is the case of a male born immediately after an eclipse of the Moon. In that case the Moon was with the Dragon's Head near the Midheaven, and the poor fellow has been haunted with visions of hanging.

Eclipses formed in the horoscopes of young people I consider most important. They come at a time when the mind will be permanently affected by events and when the character is largely in course of formation; important configurations are therefore likely to have the greatest and most permanent effects. If you look through a number of the known and published horoscopes you can easily note this. Take D. S. Windell, the assumed name of the author of the clever but comic Bank frauds that we all laughed at some years ago. Particulars of his map are given in "Notable Nativities" No. 391. There you have a South Nodal eclipse in the tenth house maturing directionally at about seven years of age. No doubt it had its more immediate results, but I am now calling attention to its effect upon the formation of character. According to all my experience such an eclipse would have the effect of dulling the moral outlook of an otherwise clever person.

Another case is that of Popejoy ("Notable Nativities" No. 771), who died as a result of ill treatment by her mistress. The eclipse occurred when she was in her third year, and the effect of it was to leave her lacking the self-protective instinct.

The late Tom Ellis, M.P., son of a tenant farmer, but subsequently Chief Whip in the House of Commons, had one which fell in the tenth house at about 14 years of age. Hindenberg also had one with the lucky Node. Mrs. Besant's early blossoming into prominence seems due to the same cause.

## Blindness

I have already referred to blindness being caused by Eclipses developing in the horoscope. At the outset I may as well say that in all cases of blindness occurring after birth, the Dragon has a substantial hand in the matter.

Much information has found its way into ordinary astrological works regarding the general indications that point to the danger of this calamity overtaking one; but owing to the Nodes not being understood there has always been much uncertainty in the matter. I have never found a case of either temporary or permanent blindness occurring without Nodal action playing a great part, and the same applies to any other serious defect of sight.

Either Light square the Nodes is detrimental to sight and liable to cause imperfect focusing. The following points in the Heavens are notorious for their evil effects upon sight:

| Pleiades | 29 Taurus 0 | 23N51 |
| Aldebaran | 8 Gemini 40 | 16N21 |
| Praesaepe | 6 Leo 7 | 20 N15 |
| North Asellus | 6 Leo 25 | 21N45 |
| South Asellus | 7 Leo 36 | 18N27 |
| Antares | 8 Sagittarius 39 | 26S16 |

One or both of the Lights conjoined with either of these points or in parallel to them is somewhat threatening to sight; so are their squares and oppositions. The middle of Virgo (about 16 degrees) acts the same, but individually these points will effect little or nothing and there must always be more than one of these afflictions. It is said that the Hindus consider that the rulers of the first house and the second in conjunction in the sixth or eighth or twelfth house is a strong or certain indication, but I do not think that any one indication is sufficient of itself. In nearly all cases the period of its occurrence will be at the

## The Moon's Nodes

time that an Eclipse or Eclipse-like formation is formed by diurnal progression in the horoscope, or the conjunction or opposition of the Sun and Moon square Dragon's Tail.

I could add a great deal more upon this subject, and especially as to indications in the radix, but do not wish to prolong this section unnecessarily, but only to give the clues that must be added to my remarks upon eclipses by directions, for the matter to be understood. I do not wish such an unfortunate want of knowledge to continue as was shown by well-known astrologers in the case of Sir C. A. Pearson. In that case all usually impugned points in the Zodiac were in action, but as that action was mainly by their declinations, they seem to have been overlooked.

But what I particularly want to call attention to is that Sir C. A. Pearson's figure is an Eclipse horoscope, and in my view one that most emphatically indicated approaching blindness. I will ask readers to refer to his horoscope at the end of this book. It can there at once be seen that the Sun has only 38 degrees to travel to meet the Dragon's Tail. In that case it is obvious that one or more bad Eclipses are ahead. What really happened was that a weak Solar Eclipse was formed by direction at 20 or 21 years of age at the undesirable end of the Dragon. At about 35, a Lunar Eclipse occurred within one and a half degrees of the Nodes, and at 50, another Solar South Nodal Eclipse! By that time he was quite blind.

## Pre-Natal Eclipses

I cannot close this chapter without adding a few remarks about Pre-natal Eclipses.

All points formed on the Ecliptic by Eclipses during the period between conception and birth, if affecting either of the true angles, have a marked effect upon the potential individual. Should the eclipse be a South Nodal one, and fall upon any point that becomes one of the angles of the Natal figure, it has a great tendency to prove fatal to the child being reared. A North Nodal Eclipse if not near a square or opposition of an important point in the after formed birth figure, would, of course, have the opposite effect. The same applies to the position which will be occupied by the Sun or Moon or the Part of Fortune at birth.

Eclipses occurring in the earlier stages of gestation seem to be of much less account than those formed about the middle of the period. Again, those formed in the latter part of the period also have a decreased effect. No forecast with regard to the possibilities of rearing a child should be made without taking this factor into consideration.

Of course, most children do have pre-natal eclipses indenting some part of the horoscope, but in many cases these fall on an unimportant part of the natal map and, therefore, would not very easily be traceable. The main effect of the eclipses seems to exhaust itself in the early years of life, when the native survives them, but no doubt they have an effect also upon the stamina of the individual. They also seem to have something of the effect of eclipses formed during childhood with regard to character and such like developments, especially those that are formed during the latter part of the period of parturition.

Astral forces cannot be accurately weighed, but here we have one of considerable consistency in their workings. When Caput is in the northern half of the signs and Cauda in its own half, the effects seem most reliable. Note especially that the signs holding the Dragon have always to be taken into account in judgment, just as you would consider the signs held by the Sun and Moon, for they have great effect upon personal characteristics, health, etc., and the Tail is apt to create a weak spot in the part governed by the sign that it occupies, and that especially so if it be in Northern signs.

# Chapter VIII

# The Lights Individually and the Nodes

THE EFFECT of one of the Lights with a Node is always strong. The Sun is especially favoured by being with the North Node, but is less fortunate if with the South Node. The Moon is greatly strengthened by being with either Node, and rendered much more fortunate provided the North Node is elevated; in fact, this is the most essential point of the Moon's dignity. If, as I stated before, the Moon is with either Node and the North Node is elevated, it adds to a person's stature and also makes him or her more fortunate in all the affairs of life. The Sun appears to increase a person's stature by being with the North Node if uppermost, but not otherwise. The Moon is always in a most fortunate position if placed with an elevated Caput. Taking a Light with a Node you have formed in the horoscope a fixed point which may eventually be joined or opposed by the fellow Light in the ordinary course of diurnal progression and so forming practically the condition of a true eclipse. Should the Sun be the Light with the Node, Eclipse-like directions must occur more than once in the course of an average lifetime. On the other hand, should it be the Moon with the Node there is a chance of a duplicate of one or other form of the real thing. Obviously after what can be seen of the great and prolonged effect of real eclipses in the horoscope, these somewhat unreal ones cannot be supposed to pass without effect. All my experience in directing vetoes that idea. There are a large number of such cases in published horoscopes. To name a few—the King of the Belgians, Vaillant the Anarchist, Queen Alexandra, Joseph Chamberlain—while Mrs. Pankhurst, the noted Suffragette, had the peculiar experience of both kinds, namely real and apparent Eclipses or Eclipse-like positions of a very unfavourable kind. Her figure is given in "Notable Nativities" No. 991. Those who are familiar with her fame and general affairs may find much food for study in these positions, and an amount of astrological information which they did not expect.

I have seen the question raised as to whether directions can alter a person's position by raising them above or placing them below what is shown in their horoscopes. That is not part of my subject now, so I will not deal with it. Why I mention it is in order to point out the necessity of considering the position of these very wide-orbed points, and allowing for the promise of good or evil that is by reason of the width of their orbs already working up in the natus. If this is overlooked—and hitherto it has been overlooked in most readings—you have an apparent disparity between the natal figure and the after developments. I do not mean that I think this covers all the cases that give rise to the question, but it certainly covers a large proportion of them.

# Chapter IX

# The Moon's Nodes in Relation to the Planets

WHEN WRITING of the interaction of the Moon's Nodes and the Planets, I am fortunate in having to deal with a proposition that was set forward by William Lilly in the middle of the 17th century. Certainly he appears only to have understood their effects, or indications, when the connection was by conjunction. As that is a simple thing to do, I will deal with them in that way first. The results of their aspects to planets is not such a simple matter; and I can understand, in an age that has found astrologers who do not see and observe the most palpable effects of the Nodes, the more obscure ones being passed over.

Zadkiel, in publishing a somewhat emasculated version of Lilly's "Introduction to Astrology," under heading of Chapter XII, gives on page 52 the following quotation:

> "The Head of the Dragon is masculine, of the nature of Jupiter and Venus, and of himself a fortune. The Tail of the Dragon by nature is quite contrary to the Head, for he is evil. I ever found the Head equivalent to either of the fortunes, and when joined with the evil planets to lessen their malevolent significations; when joined with the good to increase the good promised by them. The Tail of the Dragon I always, in my practice, found, when he was joined by the evil planets, their malice or the evil intended thereby was doubled and trebled, or extremely augmented, etc., and when he chanced to be in conjunction with any of the fortunes, who were significators in the question, though the matter by the principal significator was fairly promised and likely to be perfected in a small time, yet did there ever fall out many rubs and disturbances, much wrangling and controversy, that the business was many times given over for desperate before a perfect conclusion could be had;

## The Moon's Nodes

and unless the principal significators were angular, and well fortified with essential dignitaries, many times unexpectedly the whole matter came to nothing."

Having given this much of what Lilly says upon the effects of Caput or Cauda being with the planets, his emending imprimatur adds a footnote of his own, saying:

"These points are of no consequence in nativities, except as regards the Moon, who brings benefits when she reaches the Head in the Zodiac by directional motion, and evil when she reaches the Tail."

So apparently, according to Zadkiel, the Great Father may use the Nodes in answering our horary questions, but in nativities he only uses them when they form a conjunction of the Moon by direction!

Unfortunately I have not a full text of Lilly's writings by me, but I have that which I conceive to be of some little value—my own long years of study and observation. Lilly describes the Dragon's Head as having a power similar in nature to Jupiter and Venus combined, and so it seems to me, especially when with either Saturn, Jupiter, Mars or Venus. On the other hand the effect of the Tail with either of these planets seems as objectionable, or nearly so, as that of the Head is good. The Tail always suggests itself to my mind as well symbolized by the muddy sediment in a drain.

The question of aspects between the Nodes and the Planets is a more complicated one than appears at first sight. The question of their conjunctions, however, is absolutely simple, and I will deal with this first accordingly.

For my purpose I would divide the Planets into two classes. Saturn, Jupiter, Mars, Sun and Venus have each a moral character attached to them as part of their nature, though often obscure. Any astrologer of experience knows somewhat how to value that moral nature. the Sun and Mars individually warm and generous, the beneficence of Jupiter and the peaceful kindness of Venus when not interfered with are known factors, as also the patience and serious-mindedness of Saturn. These are moral characteristics to be affected by all the various positions in the horoscope, and it was no doubt with this in his mind that Lilly wrote some of his rather extreme statements. My observation of the effects of a Node with any of these five confirms his statements, though perhaps I would modify them slightly. Either of these planets are strengthened and improved by connection with the Head, and seem to lose their virtue and beneficence when with the Tail. Not that they become weaker or less active by the evil conjunction. It is mainly the virtue or nobility that is gone. Perhaps Saturn suffers least noticeably—the similarity of his nature with that of the Tail would account for that—but it is more powerful, vicious and more destructive to life.

Jupiter, when with the Dragon's Tail, is apt to give deformities as well as loss of his chances of honourable expansion in the realms of finance, receipt of patronage, etc. The native is liable to suffer by one or more of the diseases or accidents that these effects usually have a hand in. Of course, the house, sign, elevation, etc., must always be considered, laying special stress upon elevation, but the Planet does not seem weakened by contact with the Tail, and its aspects are decidedly stronger for the contact—by aspect most unfortunately so. The conjunction of the Dragon's Head with the Sun, or any of the Planets, obviously strengthens and enhances the good qualities of the Planet in question. As the North Node's influence is of a combined Jupiterian and Venusian nature, it adds these qualities to the Planet with which it is in conjunction, and when Saturn is in question has a modifying effect upon the seriousness of this planet. An astrologer who had taken my cue in these things informs me that he has met with several operatic singers and vocal performers generally who have had Venus with the Dragon's Head. I know that Madame Patti had Venus with the Head, and that it was also parallel Jupiter, so that here was the double effect that has puzzled some astrologers.

The second class of Planets—namely, Neptune, Uranus and Mercury—form a set by themselves. Neither of these planets can be said to have a moral nature. I know that this statement will, as far as the first named is concerned, give a bit of a shock to those who have imbibed their ideas of this planet from certain religio-philosophical sources, but practical astrologers will recognise the truth of this statement at once and accordingly I describe Neptune as feminine, rather moist, dreamy, fond of mild sensations and unmoral; Uranus as masculine, dry, positive, energetic, inquisitive and the father of inventions, unmoral; Mercury slightly masculine, active quick-thinking, unmoral. The Tail has not much to destroy with either of these, though to the first two planets at any rate, it can add pernicious qualities of its own. The Head can also supply some of its own qualities to them, namely with Neptune a fine imagination while the other end can make it more active.

Uranus is the planet that I have seen its disturbing influence make most play with. This, I suppose, from the nature of the planet would be expected. Having Caput in the same degree with Uranus myself, and several of my relatives having the planet with one end or other of the Dragon, has given me good opportunities of observing his action at close quarters, and if students like to look at the published horoscopes of people who have attracted attention, they will be surprised to find what an effective part Uranus can play when joined to the great overshadowing power of the Dragon. For many reasons there are more people who make themselves notorious who have Uranus conjunction the Dragon's Tail than those having the Head with this planet. Considering the original, inventive and unmoral nature of Uranus, this is not surprising. Take the case of the notorious Mrs. Maybrick, for instance, who in order to rid herself of an old and wealthy husband hit upon the device of abstracting poison for the purpose from flypapers! Her map shows Uranus conjunction the Dragon's Tail suitably placed for the purpose in Gemini in the tenth house.

A certain Chairman of the London County Council (now deceased) who was notorious for his capacity for undoing the work of the Progressive members of that body, had Uranus and Cauda joined in Pisces in the tenth house. For the purpose of destructive work such a conjunction is unequaled. President Garfield also had this combination.

On the other hand, we have an example of the Head conjunction Uranus in the case of General Grant, another President of the United States of America ("Notable Nativities" No. 237). It will be remembered that he was a Federal General in the Civil War of America and was largely responsible for bringing that war to a successful conclusion for his own side, and later became President.

In Raphael's Ephemeris for 1923 is given the horoscope of one styled "R. P." who had been the tutor of the then "Raphael" in his early days. Raphael's laudation of "R. P." was no doubt well deserved. Unfortunately neither of them understood the Nodes; but "R. P." had Caput in the first house occupying the same degree as Uranus, with Neptune a few degrees below them; hence the erratic but clever astrologer.

Mercury with the Nodes is an exceptionally active gentleman. This conjunction is often found in the horoscopes of writers and journalists; as far as journalists are concerned it has usually been the Tail which has been joined to the planet as far as my experience goes. "Notable Nativities" has many such cases, but of course I know nothing of these individuals personally. All of them, however, have been people of great mental capacity in their occupations.

Now as regards aspects, the parallel has perhaps the most powerful relationship between points in astrology. Always make notes of the declinations of the Nodes for any point that holds the parallel of those points is greatly affected. For the benefit of students who depend solely upon the ephemeris for their information I may as well give an easy way to find the Dragon's declination. The declination of the Nodes, or of any point upon the Ecliptic, as the Midheaven, Ascendant, or Part of Fortune, can always be readily ascertained. Turn over the ephemeris until you come to the Sun shown as occupying the same point. The declination of the Sun is given for each day at noon and only a little mental arithmetic is needed to apportion the declination of the Node to the Sun's declination at noon, and the declination of any other Ecliptic point can be obtained by the same means. When you find the declination of the Dragon's Head, the Dragon's Tail being just opposite would hold the same degree. Whichever one is north, the other must necessarily be south. For example, on August 1, 1926, the Moon's Node is given as in 15 Cancer 03; we therefore turn back to when the Sun transited that point, namely July 8, and so obtain its declination, viz., 22N36, and enter it with the declinations of the Planets and Ecliptic Points in the Map. The declination of the Tail need not be entered as it is the point exactly opposite. Anything within one degree of these figures counts as a parallel, so Venus on that day being at noon 22N32 can be described as a fairly close parallel of the Nodes and is accordingly vivified, strengthened and made more active and if Caput gives promise at the time there is all-round gain.

A matter which has most puzzled astrologers in the past and has turned many from making experiments with the Nodes is that where they may have found an aspect or direction between a planet and the Nodes act powerfully in one case at other times they found several such aspects which appeared to have no effect and have therefore abandoned the study of them. I will throw the necessary light on this difficulty, but first of all let me say that this trouble never occurs with the parallel or conjunction, but only with the apparent square or sextile and trine aspects. To ascertain whether these are a negligible quantity or a powerful aspect or direction you should remember that the Nodes are always points upon the Ecliptic; so are the Ascendant and Midheaven and, likewise, Fortuna. The Nodes can be effectively directed to these points in every case, but the planets stand in a different category. Each of these has a circle of its own at more or less obliquity to the ecliptic; accordingly when away from their

# The Moon's Nodes

own Nodes they are possessed of latitude, and unless the Moon's Nodes are near the Planet's Nodes when that Planet acquires latitude, the aspects become weak. But when the Moon's Nodes are near a Planet's Nodes, or a Planet is near its own Node, the aspects are very strong; so also if a parallel of declination exists between them. In this case the aspect gives its own nature, good or bad, to the parallel; hence when the Moon's Nodes are not near the Nodes of a Planet it is necessary always to know whether that planet has much latitude. If it has much over a half degree of latitude the aspect will be of little consequence. As these latitudes are always given in the ephemeris, it will be easy to note this, but when the Dragon's Head or Tail is near a Planet's Node, and so capable of a strong aspect whatever the latitude of the Planet, this will not be found in the ephemeris as a rule, as these points are not given there.

I accordingly append a list of the Planets' Nodes at the end of the book, which will serve for anyone's lifetime, for though they have a slight forward motion of their own, it is so slight that for this purpose the list I have given can be used for 100 years at least, either counting backward or forward. Moreover, the Moon's Node within 15 or 20 degrees of a Planet's Node will certainly make any aspect active. Uranus is never more than 48 seconds of latitude, and hence it is much more frequently noticeably susceptible to these aspects than the other major Planets.

Neptune has been spending 14 years within touch of its own North Node, and it will be about 68 years before it gets within a half degree of the ecliptic again. Then it will be approaching its South Node—a great fact for students of political astrology to take note of. (The writer has paid some heed to these facts for many years and applied them roughly to the history of the last few centuries, and hopes one day to make his conclusions public.) Its good aspects with the Dragon lift the conception and imagination; its square when strongly operative is most disastrous, and gives danger from poisoning. An accomplished London astrologer with the Node in Leo square Neptune informs me he himself was once dangerously poisoned. Another case, "Notable Nativities" No. 362, gives a mysterious death with this position. The square of the Nodes to any planet is particularly vicious whenever the conditions I have laid down exist. I do not think I need particularise further except to add that Mercury, which crosses the ecliptic nearly every month, is a possible exception. Prospective lovers should beware of the effective square to Venus, as it is almost sure to bring disappointment. The horoscope of Mrs. Besant illustrates the activities of the Dragon in many ways, and a long chapter might be written on this horoscope dealing with its many-sided effects. Without it such a career as hers would have been impossible in spite of the strength of the map, but I have no space nor desire to enlarge upon it generally now; students should study it at their leisure when they have exhausted this book. The only thing in her horoscope to which I will call attention now is the fact of Uranus conjunction Cauda being in the Ascendant and planets with Caput in the seventh house.

General Sir R. Baden-Powell's horoscope is an example of a telling square of the Nodes to Saturn. Caput with Jupiter in the second house sextile the Ascendant and trine Midheaven is a promise of success of dazzling proportions. But Saturn throwing that square to Jupiter and Caput must have been a serious drag upon the road to glory, and have made every effort a laborious one.

## Chapter X

## In Conclusion

THERE ARE various points in connection with the action of the Nodes with which I have not been able to deal in the foregoing, such as their effects in everyday life, transits, etc., but the following may be of use to students who desire, as I hope they will, to study the Nodes for themselves.

When a child is born with the Nodes in either pair of angles, do not immediately commence delineating, for with these positions you at times get the most extreme forms of abnormality without any horoscopical indications to point them out.

In other maps any deformity or striking peculiarity is easily detected, but with the Nodes angularly placed this is generally not the case. The child may be born with half the head missing, or any of the vital organs missing or deformed; or it may be a microcephalic idiot, though the horoscope shows a philosopher or an astute man of business. "Notable Nativities" has many such cases. If the compiler of that valuable little reference book had included the Nodes, I am sure he would have noted this peculiarity, and also some others, and so the Nodes might have received his valuable attention. In that case, perhaps, this little book would have been unnecessary.

Where the Nodal deformities occur they are usually of a very pronounced and observable kind, or such as an experienced nurse would readily notice, and when she is satisfied it is time to proceed; but to read such a horoscope without first enquiring is, when the Nodes are so placed, to invite ridicule, or perhaps to fall into a trap.

Since writing this last paragraph the news has been published of the birth of a Royal Princess. To this birth the above remarks are directly applicable. The time of birth is given as 2:40 a.m. (Summer Time), April 21, 1926. I therefore erect a figure for 1:40 a.m. G.M.T. for that date. The figure shows the twenty-second degree of the sign Capricorn rising, with the Dragon's Tail one degree above it. This latter position precludes any judgment being given upon the Nativity at present, for reasons stated above. There is, of course, the temptation to astrologers writing for the public dutifully to try their skill in the matter of a birth that may one day prove of first class importance to the State; and unfortunately the Nodes are not yet understood by most of such writers. But I hope my readers will loyally say nothing in this case at present, but be warned by Cauda's rising position. There is only one remark upon this horoscope that I consider opportune: the Moon's Nodes being close to Saturn's Nodes makes Caput trine Saturn a very effective aspect.

The Tail with the Part of Fortune is a somewhat dangerous position during infancy; also at any time when the Hyleg becomes afflicted. The reasons for this may not seem very apparent, but it is part of a great and important subject that I hope one day to introduce to the astrological student.

## The Nodes in Everyday Life

The student who turns over his ephemeris and who scans the Aspectarian and (as must usually be the case) overlooks the Dragon, misses the indications that make or mar any scheme or action that falls under its influence. In all elections of time for the commencement of anything, or for entering into any bond or undertaking, the good auspices of this power are of the utmost importance. It can mar the (otherwise) best selection, but with the blessing of Caput not much evil can come. Of course, Caput on the cusp of the Midheaven is usually the best testimony;

Cauda there, the worst. But look at its relationships all round.

As an example from the launching of ships, the unfortunate "Kerangie" ("Notable Nativities" No. 852) was launched with the Nodes square the Midheaven. It had an accident upon its trial trip and struck a rock some months later.

The founding of the Theosophical Society is worth careful note. Inaugural meeting held in New York, November 17, 1875 ("Notable Nativities" No. 244). The Planets being all but one in fixed signs might be held to account for its continued existence; but what a figure to exist under! Without the saving grace of Caput in the tenth house in effective trine to Venus, and with the Moon applying to the trine of it, and so casting an air of respectability over its many vicissitudes and ensuring the support of good lady helpers, what sort of a reputation would it have piled up? The laying of the Foundation Stone of the new Headquarters of the Theosophical Society at Euston is given in "Notable Nativities" No. 936. This figure without the Nodes seems to have much more promise, but Cauda is five and a half degrees above the Ascendant and only six and a half degrees from the rising Jupiter, whose exact sextile to the Sun should have promised so much. Astrologers familiar with the project will be able to judge whether my contention as to Nodal effects are borne out in this case. I will not multiply words nor quote cases unnecessarily. In all your undertakings, be sure the Dragon is not against you. You can tide over much with his beneficence in your favour.

## Transits

The Nodes act as transits, as and under the same conditions as do the Planets.

All Eclipses that occur as transits regard in the same way. Never mind the panicmongers who make so much ado about them. They can only act upon the directions formed, and in so doing are very powerful and will prove good or bad according to their nature and aspects; and their power as transits seems to be pretty well exhausted by the time that another eclipse could be formed at the same Node—that is, within a little less than a year.

## Apologia

In concluding this attempt to make something known of this great overshadowing power, it is with the knowledge that a great deal more is to be discovered of the details of its influence. I have included no guesswork and no speculations; and have left out all matters upon which I am uncertain. There are things that I would have liked to have asked students to direct their attention to, but thought it best to keep to what I could speak positively upon. It may, however, be of interest to my readers to know that my experience in this connection commenced in 1877.

The examples I have quoted have been publicly known cases, and that without much selection. I could easily have given examples from my own private collection of horoscopes, which I think would have been better for my purpose, as they are cases of which I have some personal knowledge, but I thought it best to rely mainly upon published cases which the student could easily refer to.

There are phases of Nodal action that could not be explained in this book, but if I find this attempt is appreciated it is my intention to introduce and let light in upon what is behind Fortunae. That I believe to have been a close-kept secret for ages. With that part of astrology understood, the Nodes would be of much greater interest.

This book has been written under stress of other business, hence the reader may find much to condone. I would have liked to have put off its production until I had more time to devote to its preparation; but age cannot advisably put things off indefinitely. Others, no doubt, have had the knowledge and have let it die with them. I regard knowledge as a Sacred Trust and not for personal use only; hence this book.

## The Moon's Nodes

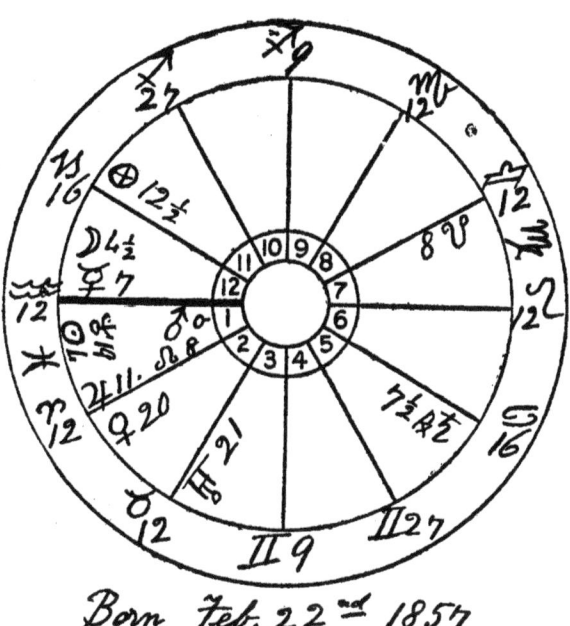

Gen. Sir Baden-Powell
(From Not- Nat- 837)

Born Feb. 22nd 1857

| | ☉ | ☽ | ♅ | ♆ | ♄ | ♃ | ♂ | ♀ | ☿ | ⊕ | ☊ | Asc | M.C |
|---|---|---|---|---|---|---|---|---|---|---|---|---|---|
| LAT | — | 4.34 | 1-7 | 0.13 | 0.26 | 1.7 | 0.36 | 1.37 | 0.28 | — | — | — | — |
| DEC | 10.41 | 23.40 | 5.12 | 17.52 | 22.49 | 3.15 | 0.35 | 9.16 | 18.8 | 22.53 | 3.10 | 17.10 | 24.50 |

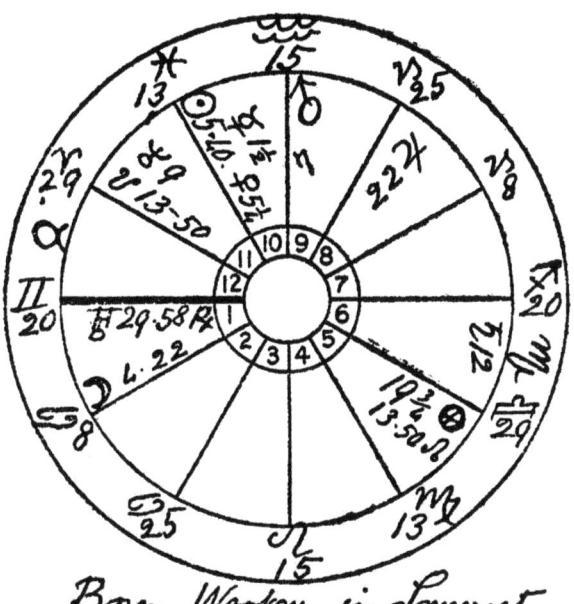

Sir C. A. Pearson

Born Wookey in Somerset
Feb. 24th 1866 at 11.0.a.m.

| | ☉ | ☽ | ♆ | ♄ | ☊ | ♃ | ♂ | ♀ | ☿ | ⊗ | ☊ | Asc | MC |
|---|---|---|---|---|---|---|---|---|---|---|---|---|---|
| LAT | — | 5·9 | 1·30 | 0·16 | 2·32 | 0·7 | 1-1 | 1·26 | 2·3 | — | — | — | — |
| DEC | 9·26 | 18- | 2·11 | 23·45 | 13·8 | 21·41 | 19·2 | 10·26 | 12·6 | 22·5 | 3·12 | 17·10 | 21·50 |

# The Moon's Nodes

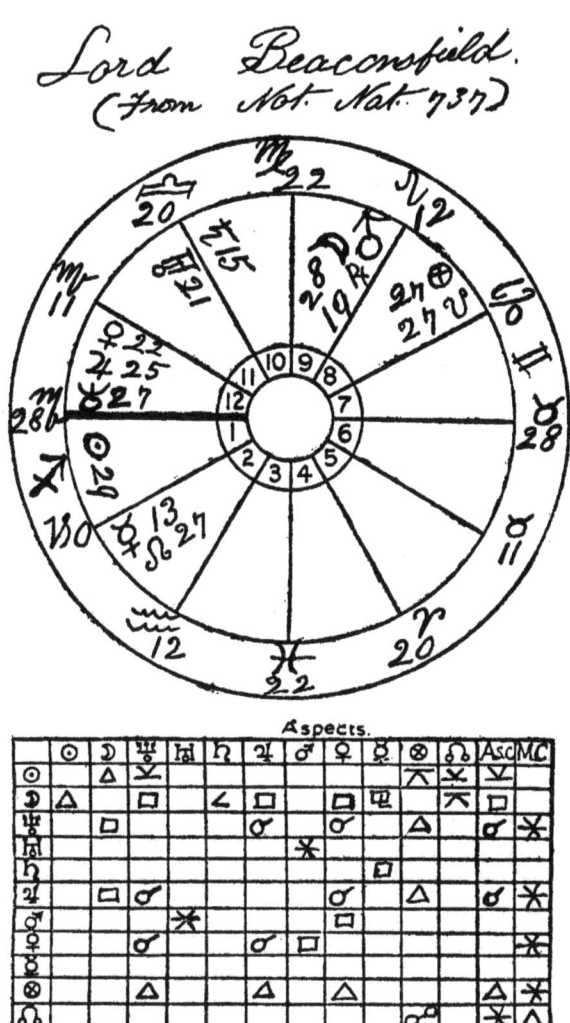

## Position of the Planets' Nodes

| | |
|---|---|
| Neptune | 10 Leo 27 |
| Uranus | 13 Gemini 22 |
| Saturn | 23 Cancer 0 |
| Jupiter | 9 Cancer 26 |
| Mars | 11 Taurus 56 |

These have a slight motion in the order of the signs, but so slight that it would take centuries for these motions to make any material differences.

With the minor Planets (Venus and Mercury) we need to note in the ephemeris the times of their crossing the Ecliptic, as it is the geocentric position of these crossing points that matter. To do this, note under heading of Latitude when either of these Planets crosses from north Latitude to south Latitude, or vice versa, and its longitude, for that day will do for its apparent Nodal position. With Mercury this occurs every few weeks, and whenever found within about a half degree of latitude, must be regarded as capable of receiving strongly effective aspects from the Moon's Nodes, as explained earlier.

## How to Find the Position of the Dragon's Head at Any Date

The Node moves backwards through the signs at the rate of three minutes per day or 19°20' per year. It makes the complete journey round the Ecliptic in 18 years and 223 days.

On January 1, 1926, its longitude was 26 Cancer 17, so if we deduct its movement for a year we find it will be 6 Cancer 57 on January 1 following. To go back a year add the 19°20' and we find that on January 1, 1925, it was 15 Leo 37.

For long periods the following table will be useful:

| Motion in Years | Equals | Signs | Degrees |
|---|---|---|---|
| 5 |  | 3 | 6:40 |
| 10 |  | 6 | 13:23 |
| 15 |  | 9 | 20:03 |

|     |                |   |       |
|-----|----------------|---|-------|
| 20  | The Circle plus | 0 | 26:50 |
| 50  |                | 8 | 07:04 |
| 100 |                | 4 | 14:05 |

Always remember in calculating to a future date to deduct. For a previous date, add. The error (if any) will not exceed a few minutes.

www.ingramcontent.com/pod-product-compliance
Ingram Content Group UK Ltd.
Pitfield, Milton Keynes, MK11 3LW, UK
UKHW041422180426
11947UKWH00007B/247